# Sound

## Contents

# Revision

Did you know . . . ?

Things which vibrate very quickly are sources of sound.

Some materials let sound through them. Some materials do not let sound through them. Some materials reflect sound.

We hear things when sound reaches our ears.

**5** What can you hear?

**3** What can she hear?

**1** What can they hear?

**6** What can you hear now?

**4** What can she hear now?

**2** What can they hear now?

**Task 1** **What can you hear?**

⬧ Look at the pictures.

⬧ What can be heard? Why?

⬧ Draw and write to show your ideas. ③

⭐ **We can change the pitch of a sound. The faster the vibration, the higher the pitch.**

The **pitch** of a sound is how many times something vibrates in a second.

**you need:**

- a rectangular tissue box

- different sized elastic bands

- a pencil

Task **2** A box guitar
...................

✦ Make a guitar like this one.

✦ Make changes so that it makes three different notes: a high note, a low note and one in between the two.

You have changed the **pitch**. The pitch of a note is how high or low the sound is.

✦ Find three different ways of changing the pitch of the note.

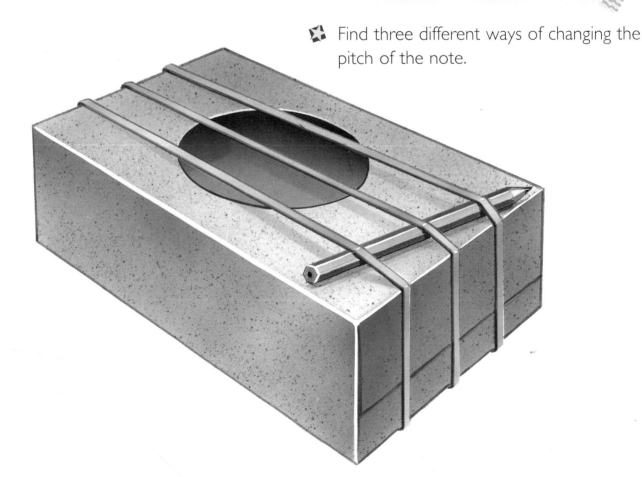

# Fact File

# The pitch of a note

The more times something vibrates each second, the higher the note. The fewer times it vibrates each second, the lower the note. Pitch is measured by how many vibrations there are in a second (vibrations per second).

## Task 3 Investigation
### Investigating pitch

PCM 2

✷ Investigate the pitch of your box guitar, or of a stringed instrument like a guitar or a violin.
How does a change affect the pitch of a note?

✷ You could change:
- the length of the string you test
- the thickness of the string you test
- the material it is made from.

✷ Predict what will happen before you do your test.

✷ You could observe whether the note went up or down in pitch.

✷ You must try to keep the **volume** (the loudness of the note) the same.

PLANNING BoARD

Our question _____
We will change _____
We will measure _____
We will keep these things the same to make our test fair _____
This is the table we will use. (Put in the headings. Fill in the left-hand column.)
We will use these things _____

| What I changed | What I predicted | What I observed |
|---|---|---|
|  |  |  |

✷ When you have finished your investigation, use the table for your results.

✷ Complete these two sentences which explain how to change the pitch of a note.

To make a note with a higher pitch, you should change a string by . . .
To make a note with a lower pitch, you should change a string by . . .

✷ Explore a tuned percussion instrument like a xylophone or chime bars. How is the pitch of the note changed?

# Fact File

## Twanging rulers, moving air

When you twang a ruler on a table edge, you move the air around it.

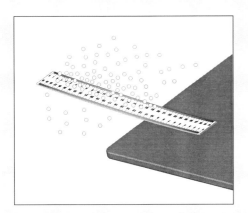

① As the ruler moves upwards, it squashes the air together above it, thinning the air below.

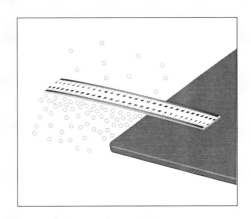

② As the ruler moves downwards, it squashes the air together below it, thinning the air above.

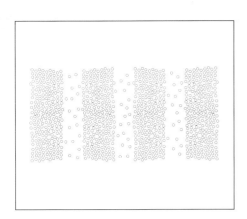

③ As the ruler bends back and forth, it squeezes the air.

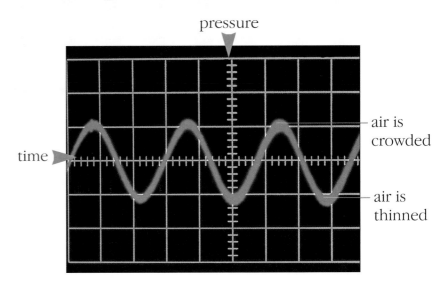

pressure

time

air is crowded

air is thinned

④ The sound moves out from the ruler through the air. These changes are called **waves**. The **crest** of a wave is where the air is crowded or squashed; the **trough** of a wave is where the air is thinned.

The air isn't rushing around the room. Tiny particles of air are pressing on their neighbours, like people pushing each other in a crowd.

All musical instruments work b squashing the air in this way.

Sound travels through air at 330 metres per second.

6

## Task 4  Balloon music

 Blow up a party balloon. Hold the neck closed between your finger and thumb. Let the air out gradually so that the balloon plays a note.

 Change the pitch of the note.

Play a tune.

## Task 5  Bottle flutes

PCM 3

Put a bottle on the table and hold it steady.

Rest your lower lip on the edge of the opening and blow. The bottle will make a hooting sound.

Get some bottles in different sizes and blow across each one.

What do you notice?
Write one sentence which links the size of the bottle to the pitch of the note.

Get several bottles the same size, but fill each with a different amount of water.

Blow across each of the bottles.

What do you notice?
Write one sentence to explain what happens when the amount of water is changed.

### You need:

- Clean, empty glass bottles, some in different sizes and some in the same size
- Water

 **Safety point:**
Take care when using glass bottles.

a penny whistle

## Task 6 · Exploring wind instruments

Many musical instruments make a note when you blow through them.

In many brass instruments, your lips vibrate.

tuba

trombone

clarinet

bassoon

In many woodwind instruments, a thin reed vibrates.

✡ Explore a wind instrument like a recorder, penny whistle, flute, bugle or trumpet.
What vibrates when you blow them?
How do you make different notes with them?

✡ Look at these photos of different wind instruments. What vibrates when you blow them?
How do they makes sounds of different pitches?

PCM 4

✡ Choose a musical instrument.
Tell the story of how it makes sounds.

## Task 7 · A pig grunter

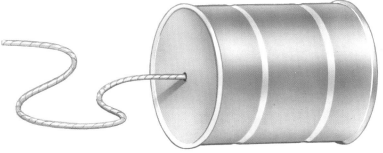

### You need:

• a large, clean empty can with no sharp edges

• a hammer

• a long nail

• a paper-clip

• 1 m of thick, stiff string, either nylon or waxed

✡ Make a hole in a clean can. Knot a thick string in the hole.

✡ Slide your index finger and thumbnail down the taut string.
What is vibrating? How does the can **increase** the vibrations? How can you vary the sounds?

# Fact File

## Different types of instrument

People have always wanted to make music. Flutes from ancient times have been found in China and there are pictures of musicians on Egyptian tombs. Musical instruments from every age and every country can be divided into four types: **idiophones, membranophones, cordophones** and **aerophones**.

Gongs, bells and rattles are called **idiophones**. In xylophones and this thumb piano, the sound is increased because of the box.

Drums and tambourines are called **membranophones**. A stretched skin is struck with the hand or with a drumstick.

In this African drum, the drummer can squeeze the laces to change the tightness of the skin and the pitch of the note.

Stringed instruments – harps, lutes, lyres, zithers, guitars and fiddles – are called **cordophones**. Making the string longer or shorter, tighter or looser, changes the pitch.

**Aerophones** are horns, pipes, trumpets and flutes.

The digeridoo is like a trumpet and is used by Aboriginal Australians to produce a low droning sound.

**We can change the loudness of sound. The bigger the vibration, the louder the sound.**

Most things that make a loud noise are either:

- hit very hard – the **force** with which a drum is hit makes it loud

- **big** – a big drum makes a louder sound than a small one when hit in exactly the same way

- made from the right **materials** – steel guitar strings sound louder than nylon ones when plucked in the same way.

Task **8** Making a loud noise

 Press your hands together very hard. You may be pressing hard, but there is no sound.

Now clap with your fingers as hard as you can. You made some noise.

Now clap with your whole hand. As long as you clap fast, you will squeeze the air between your hands and make a **pressure wave**.
Clapping hard is one way to make a loud noise.

Now list other ways of making a very loud noise.

| Loud noise | Force | Size | Material |
|---|---|---|---|
| Drum beat | Hit with a beater | Large | Plastic drum head |
| | | | |

Explosions sound loud because they make a huge pressure wave.

# Fact File

## Tuning forks

A tuning fork is used by musicians to make sure that their instrument is making a note at exactly the right pitch. Tuning forks have two prongs and are made of steel. Some have a thumb notch at the base to help you to hold them properly.

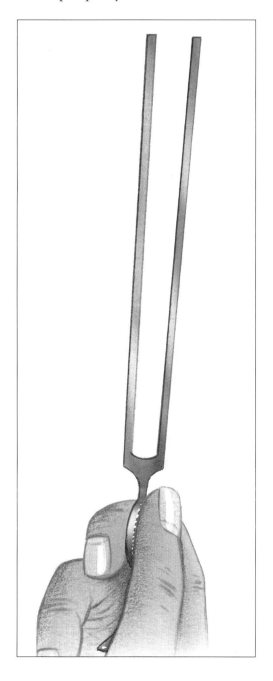

### Task 9 | Investigation
## Trying different tuning forks

 **PCM 6**

✦ Make a sound with one tuning fork. Hold it loosely at the thumb notch with a finger and thumb. Strike it on a firm – not hard – surface, like a book.

✦ Still holding up the fork by the thumb notch, hold it towards the window or the light. Can you see the prongs vibrate? Can you feel them vibrate?

✦ Now **amplify** – make louder – the sound of the fork. Strike the fork again and hold the base against a table top. Listen to the note.

✦ Now look carefully at three or more different tuning forks. Record the data you collect on a table like this.

| Letter on fork | Length of prongs | Weight of fork | Pitch - high or low? |
|---|---|---|---|
|  |  |  |  |
|  |  |  |  |

✦ Write one sentence about the size of tuning forks and the sound they produce.

# Fact File

## Wave shapes

You will remember (see page 6) that the changes in pressure that make sounds can be shown as **waves**. ▶

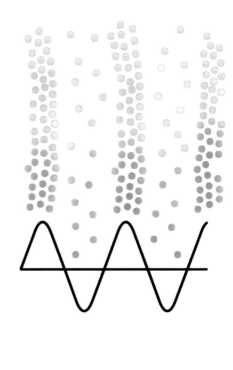

## Loudness

The volume of a sound can change.

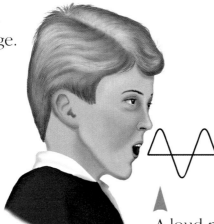

A soft noise, like a whisper, can be shown by small waves.

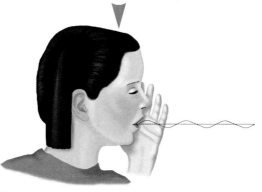

A loud noise, like a shout, can be shown by big waves.

## Pitch

The pitch of a sound can change too.

A bat's squeak can be very high. There are more waves per second.

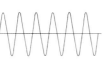

A loudspeaker hum can be very low. There are fewer waves per second.

Put a vibrating tuning fork in water.

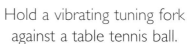

Tuning forks can do remarkable things!

Hold a vibrating tuning fork against a table tennis ball.

## you need:

- A team of Star Investigators

- tuning forks

- some different things to test the forks with

✶ How can you use this knowledge to find a way to judge the volume of sound from a tuning fork?

✶ What could you change?

✶ What will you observe or measure?

✶ What will you keep the same?

✶ What will you record?

### PLANNING BOARD

Our question _____
We will change _____
We will measure _____

We will keep _____
these things _____
the same to _____
make our test _____
fair _____

This is the
table we
will use.
(Put in the
headings.
Fill in the
left-hand
column.)

We will use _____
these things _____

| What I did with the tuning fork | What I observed or measured |
| --- | --- |
|  |  |

✶ Record your results. Explain them to your friends and to your teacher.

Let's take the school bell

or a drum

or a whistle

or a trumpet

# Fact File

## Calling for Help!

The international signal for distress is three clear blasts on a whistle, then a pause, then three more, and so on. The pause is important. The people in trouble need to listen for an answering signal.

## Task 11 · Alarm call!

Class 6 are going on a mountain walk. They are thinking about what they must do to keep safe.

'We might need to call for help,' said their teacher. 'Is shouting good enough? Or should we take something that makes a noise?'

✤ How could they find out which would be heard at the greatest distance?

✤ Would their noise-maker be heard in every direction?

✤ Which would get the attention of rescuers? Why?

✤ Which would be easiest to carry?

✤ Draw or write to show your ideas.

✤ At what time of day, and in what sort of weather is a sound signal important?

✤ How could you explore how far a whistle can be heard in different weather?

✤ Write down your ideas.

◆ Dennis is learning the piano. His teacher is desperate. Every time she says, 'Play softly, Dennis', he plays low notes. When she says, 'Play louder, Dennis', he plays high notes.

◆ Think of a way of teaching Dennis the difference between high and low, loud and soft. Write or draw your ideas.

◆ For each type of sound, draw what the sound waves look like.

# Fact File

# Making a noise

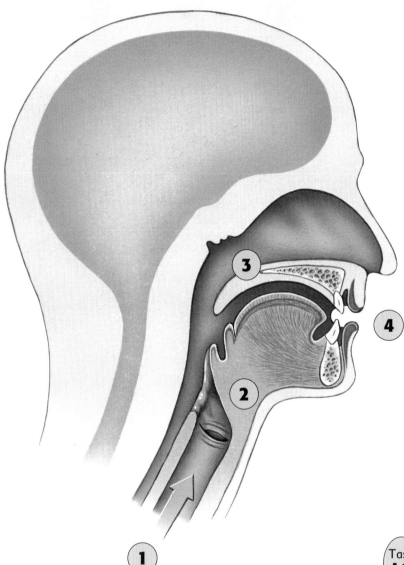

We make sounds with our vocal cords and hear them when the sounds reach our ears. Our ears are receivers for sounds, but the part that we call our ear, the ear flap, is only a collector of sounds. The sounds are received and understood inside our heads.

The picture shows how we make sound.

① Air from our lungs travels up to our mouth.

② The air passes through our voice box which has two muscular flaps. These are called the **vocal cords**. Normally these flaps are slack and the air passes between them without a sound. When our muscles pull them tight, they make the air vibrate which makes a sound.

③ The roof of our mouth amplifies the sound.

④ Our tongue, lips and teeth change the sounds to produce speech.

## Task 12  Try making sounds

 Try saying 'thin' and 'then'. Notice how you move your tongue and mouth as you say the 'th'. It is written the same way but it sounds different.

Try humming. Put your fingers on your voice box. Can you feel the vibrations?

Try making some different sounds.

# How do your ears work?

Complete this table.

| Part of the ear | What does it do? |
|---|---|
|  |  |

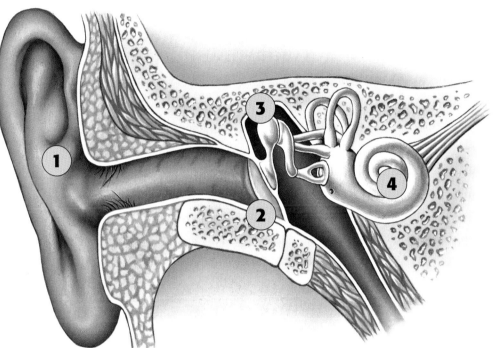

① The outer ear collects the sounds.

② The ear drum receives the vibrations. It vibrates when sounds hit it.

③ The three ear bones pass the sound along.

④ The inner ear has fluid in it. Sound vibrations move the fluid which move tiny hairs. These send messages through the nerves to your brain.

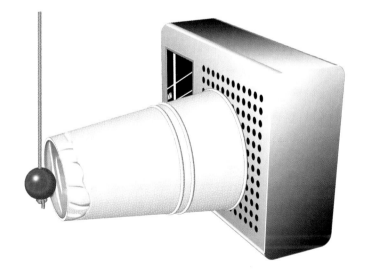

**Hint:** Glue the clingfilm to the cup. Use a small polystyrene ball.

✻ Make some paper cones. Use them as 'artificial ears'. Do they help to collect sounds?

Task
**14** Using an ear drum model

✻ Make a model of the ear drum like the one shown to understand how the ear drum passes on sound.

✻ Make a noise or play a radio close to your model.

✻ Watch what happens to the ball.

## you need:

- a sound source
- a blindfold
- a metre rule or tape measure

PCM 9

✪ How could you explore whether you need two ears to tell the direction of sounds? How could you identify the direction that sounds are coming from?

✪ What will you change?
  - The direction the sound is coming from?

✪ What will you keep the same?
  - The sound source
  - The distance from the sound
  - The volume of the sound
  - The person you are testing?

✪ What will you observe or measure?
  - How many times the sound is correctly spotted?

✪ How will you record your results? You could try a circular bar chart like this one.

**PLANNING BOARD**

Our question _____
We will change _____
We will measure _____

We will keep
these things
the same to
make our test
fair _____

This is the
table we
will use.
(Put in the
headings.
Fill in the
left-hand
column.)

We will use _____
these things _____

The number of blocks show how well Ben heard sounds in each direction.

# Only one ear

A group of children talked about what they had been doing in Task 15. They asked these questions:

> **How often should we make the sound in each position?**

> **Does it matter how much time there is between each sound?**

> **How close should the listener be to the sound?**

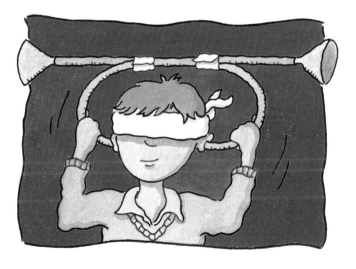

> **Can the listener hear the sound-maker moving?**

> **Does the room we use for the test matter?**

They decided to try to answer these questions:

- Does sound reach one ear a fraction of a second before the other?

- Could we do just as well with one ear?

✴ The children tried blocking one ear with a soft pad under the blindfold and repeating the investigation in Task 15. Try this.

✴ They tried the test with the listener resting the side of their head on a soft pad on the table so that one ear faced upwards. Try this.

▼ They tried this amazing device to confuse the listener. Why didn't it confuse them?

⚠ Safety point: Do not put objects straight into your ears.

★ **Some sounds can cause damage. Sounds can be muffled.**

# Schoolboy may have died while listening to the stereo

A schoolboy who cycled into the path of a car and was killed may have been distracted by a personal stereo.

Malcolm Green (13) had been warned never to use the stereo when cycling, an inquest was told. Several witnesses said that Malcolm did not appear to hear the car. PC Mervyn Brown said that the earpiece of the stereo was in the boy's ears, and the play button was on. The coroner recorded a verdict of accidental death.

# Loud sounds can damage hearing

James Hetfield is a musician with the rock band, Metallica. After months of pain in his ears, he went to a doctor who told him, 'Stop playing or you will lose your hearing.'

Hetfield explained how two or three hours of concert noise every night had left him with a ringing in his ears. Now he wears specially-made earplugs on stage. He says, 'They help me enormously, but all I can hear is grunting noises in my head which gets very scary.'

# Our ears are important

✦ Your hearing is precious. When you prevent your ears working properly you put yourself in danger.

Wearing your hood up can stop you hearing traffic.

✦ But when you are working in a noisy place, you need ear protectors to stop your ears from becoming damaged.

✦ There are strict rules about wearing ear protectors. How are these people breaking the rules? Why will the ear protectors not protect their ears? ▶

✦ Who should wear ear protectors? Name six noisy jobs.
If ear protectors are correctly worn, they can protect your ears from damage. But which materials will make good ear protectors?
Write down your ideas.

✦ Become a sound buster! Make a poster, a leaflet or a short play about protecting your ears or about some of the dangers to ears.

## you need:

- two boxes, one large, one small

- a transistor radio, which fits into the small box, or you could make your own buzzer circuit

- a range of material which might deaden sound – newspaper, bubble-wrap or polystyrene chips

- a sound sensor

PLANNING BOARD

Our question _____
We will change _____
We will measure _____

We will keep _____
these things _____
the same to _____
make our test _____
fair _____

This is the table we will use. (Put in the headings. Fill in the left-hand column.)

We will use _____
these things _____

 PCM 11

You work for SOUNDBUSTERS LTD.
A fax has arrived for you.
It asks this question.

◆ Which materials are the best sound insulators?

◆ So, which materials deaden sound best? You and your team of SOUNDBUSTERS are going to find out!

◆ Here are your clues.

Put the sound source in the little box.
Put the little box in the big box.
Close them both.

◆ You must decide:

① How you are going to pack the space between the two boxes.

② What materials you could use.

③ How you could make sure that the materials are tested properly.

④ How you will decide how well they work.

⑤ How you can show your results on a graph.

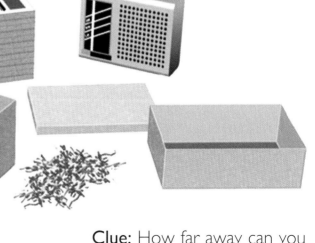

**Clue:** How far away can you hear your sound source?

# Fact File

## Stopping sound waves

Sound waves move the air. Soundproofing works because it absorbs the movement of air before it reaches our ears.

## Now try this

### Task 19 Recording your results

You need to decide:

- How many times you should repeat your test
- How to record your results
- How to present your results.

Show your results on a graph.
Will your graph be a bar chart or a line graph? Can you show the average result?

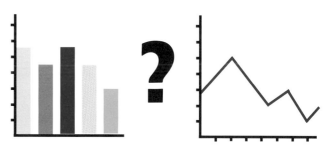

Explain your results.
Which is the best soundproofing material? Why?

### Task 20 Investigating sound control again

Look at this picture. How could this model be used to compare the soundproofing of different materials?

Try it.

### you need:

- a large box and a smaller box
- insulating materials – newspaper, bubble-wrap or polystyrene chips
- a tape recorder with a separate microphone
- a transistor radio or other sound source

# Fact File

# Measuring loudness

The loudness of sound is measured in **decibels**. The symbol for the decibel is **dB**. Here are some sounds in decibels.

a whisper, 30 dB

talking, 50 – 70 dB

rustling leaves, 0 dB

pneumatic drill, 100 – 110 dB

thunderclap, 100 – 110 dB

aircraft at take-off, 140 dB

An aeroplane, at 140 dB, is much much louder than talking, at 70 dB. It is a noise ten million times louder to the ear than rustling leaves!

**you need:**

- a portable cassette player
- a pre-recorded cassette
- a tape measure

## Task 21 The noisiest place

Sally and Robert were arguing.

**The playground is the noisiest place I know.**

**No it's not. It's much noisier in the street outside.**

They decided to test both places and other places around the school.

'You'll need these,' said their teacher. He gave them a portable cassette player, a music tape and a tape measure.
'What are these for?' said Sally.
'Switch on the tape recorder,' said Mr Bailey. He walked away down the corridor.
'I can still hear it!' he said. He walked further. 'Now I can't hear it.'
'Shall I turn it up?' said Robert.
'Oh no,' said Mr Bailey. 'That wouldn't be fair. It's time to do some measuring...'
'I see!' said Robert.
'I don't' said Sally.
'Listen,' said Robert. 'I'll explain'.

✵ What did Robert explain? How did Sally and Robert compare different places?

'I'll come with you to test the roadside,' said Mr Bailey.
'I want to make sure you are safe.'

✵ Use Photocopy Master 12 to carry out some tests of your own.

(25)

# Fact File

## Living with hearing loss

As we grow older, our hearing becomes less sensitive. We may not be able to hear very high-pitched sounds, for example. But some people experience hearing loss that is far worse. They may be hard of hearing. A few are **profoundly** (completely) deaf. Many will not live in silence but will hear a ringing or buzzing in their ears. They have a disease called tinnitus. There are 10 million people in the UK with some hearing loss.

Organizations like the Royal National Institute for the Deaf (RNID) produce aids to help people with hearing difficulties to live normal lives.

The photos show just some of them.

Walkabout - the lights show the baby is crying

Textphone - type in a message and read one back

Clock - vibrating pad goes under a pillow

Hearing aid - a microphone picks up sound and and amplifies it

Smoke alarm - vibrates under the mattress

### Task 22 Make a hearing device

✦ Design and make a device to help someone with hearing difficulties.

✦ Think first about the everyday things which they might find difficult.

# Fact File

## Communicating without sound

People who have difficulty in hearing can communicate in a variety of ways.

### Hearing aids

The first hearing aids were sea shells or animal horns with the ends cut off. When radio valves were invented, it was possible to make much more powerful hearing aids but they were the size of a small suitcase. Modern aids use transistors and are tiny, but they do not make hearing normal. They amplify **all** the sounds in the environment.

### Lip-reading

Some people lip-read. They watch the speaker's lips and see the words he is forming. But they must be facing the speaker who must be speaking slowly and clearly.

### Sign language

There is a whole alphabet that can be spelled out on the fingers. A much quicker way is to make signs for whole words like the children here. The signs are made as movements, they are not still.

### Total communication

Many people use lip-reading, sign language and speech together. This is called **total communication**.

'morning'

'name'

'thank you'

'boy'

'I'

'please'

'my'

'girl'

'bike'

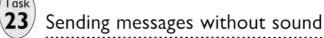

## Task 23 — Sending messages without sound

- Think of three ways that you can help a person with hearing difficulties to understand you.

- Send a simple message to a friend without using sound or writing! You could use lip-reading and sign language.

- What other soundless system could you use?

## Task 24 — Trying different echoes

The rule for good echoes is to make a short, sharp sound at least 20 m away from a smooth, hard surface like a large wall or a rock face.

PCM 14

- How do you make an echo?

- How do you think echoes work?

- Where have you heard good echoes?

- How could you make an echo in school?

- Now try investigating echoes:

  - What could you change?

  - What could you observe?

  - What must you keep the same?

### you need:

- a ticking clock or timer
- two parcel tubes
- a hard wall

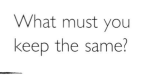

# Fact File

## Echolocation

Bats detect moths by using echoes. The bat's squeak is reflected from the moth. The bat hears it and turns towards the moth. The bat squeaks faster and faster. Each squeak tells it where the moth is.

# Can you hear through solid objects?

You can hear sound in a solid wall.

You can hear sound in a solid floor.

When the door is closed, most of the sound in the room is reflected back into the room.

# Fact File

## The speed of sound in water

Remember that sound travels better and faster through water than through air. People working on ships can use this knowledge to measure the depth of water and even detect shoals of fish.

A sound wave made by the ship under water will bounce back to it at nearly 1 500 metres per second.

 ## Task 25 The quality of sound

What makes some sounds attractive and restful, and other sounds annoying and even painful? It isn't just their volume.

Some people hate the scratch of chalk or a finger nail on blackboard. Perhaps this is because the sound is unexpected - you don't know where the sound is going next.

❖ Most people like music, although they may not agree on which sort! Why is this?

❖ Draw or write to show your ideas.

## Task 26 What sounds do you like?

❖ Use a pre-recorded tape of different sounds to investigate your friends' likes and dislikes.

❖ Try to include something scratchy and something smooth.

❖ How could you present your results on a graph?

# Checkpoint

Task **27**

## Musical instruments

PCM 15

Look at these musical instruments. On Photocopy Master 15, next to the picture of each instrument, write about all the ways you could:

- change its **pitch**
- change its **volume**.

> This special room has walls covered in fibreglass wedges. It is used for testing audio equipment. Explain how it might work.

PCM 16
What is happening in this picture? Explain what the blankets are for, and how they might work.

These saucers at the Royal Albert Hall are designed to absorb echoes. Explain why that might improve the sound of music played in the hall.